MILITARY HELICOPTERS OF THE WEST

D1809275

Front cover illustration:
Following what has become almost a tradition, Westland have now secured an agreement with Sikorsky for the licence production of the Blackhawk helicopter which will be produced at Yeovil. Export customers have yet to materialize in large number, and whether the RAF eventually purchases the type remains to be seen. However, there is doubtless a long future ahead for the Blackhawk. Illustrated is a Westland demonstrator aircraft carrying 4 × 19-tube 2.75in rocket launchers, and 2 × 20mm Giat cannon. [*Westland*]

Back cover illustration:
One of the most significant modern helicopter designs, the McDonnell Douglas (Hughes) AH-64 Apache is the United States Army's primary anti-tank helicopter. The US Army is expected to receive in excess of 800 Apaches eventually, and large numbers of AH-64s were deployed to the Middle East in 1991 during the Gulf War, taking part in hundreds of attacks against Iraqi tanks and AFVs. Engines are two General Electric T700-GG-701 turboshafts. Maximum range with internal fuel is 300 miles. [*McDonnell Douglas*]

1. F-BROK is Aérospatiale's SA.330L Puma demonstrator aircraft. The 330L was the final development model of the basic Puma design, fitted with composite rotor blades, and capable of carrying a greater payload than earlier Pumas. The Puma is in widespread service around the world, notably with France and the United Kingdom, the RAF deploying a large number of Pumas to the Middle East during the 1991 Gulf War. Interestingly, Iraq also owns 22 of the type. [*Aérospatiale*]

MILITARY HELICOPTERS OF THE WEST

TIM LAMING

ARMS AND
ARMOUR

Arms and Armour Press
A Cassell Imprint
Villiers House, 41–47 Strand, London WC2N 5JE.

Distributed in the USA by Sterling Publishing Co. Inc., 387 Park Avenue South, New York, NY 10016-8810.

Distributed in Australia by Capricorn Link (Australia) Pty. Ltd, P.O. Box 665, Lane Cove, New South Wales 2066.

British Library Cataloguing in Publication Data
Laming, Tim
Military helicopters of the west.
1. Military aircraft. Helicopters.
I. Title
623.746047
ISBN 1-85409-111-5

Designed and edited by DAG Publications Ltd. Designed by David Gibbons; edited by David Dorrell; layout by Anthony A. Evans; typeset by Ronset Typesetters, Darwen, Lancashire; camerawork by M&E Reproductions, North Fambridge, Essex; printed and bound in Great Britain by The Alden Press, Oxford.

2. The Super Puma (AS.332) was developed by Aérospatiale as an improved version of the original Puma design. With a better payload and performance, the AS.332 features a new high energy-absorption landing gear, damage-resistant transmission, and composite rotor blades. First flown in 1978, the Super Puma is in service with a variety of countries including Argentina, Chile, Iraq, Kuwait, Singapore, Switzerland and France. Although the Puma was co-produced by Westland in the UK, the Super Puma has remained a purely French venture. [*Aérospatiale*]

INTRODUCTION

It is fair to say that some aircraft types attract a much greater degree of public interest than others. For whatever reason, the media will almost invariably depict a 'military aircraft' as being a fighter or a bomber, armed to the teeth with bombs and missiles. However, the more serious observer will note that there is much more to military air operations than just air defence and ground-attack. A modern air force is also equipped with relatively large numbers of transport aircraft, aerial refuelling tankers, maritime patrol aircraft, trainers, electronic counter-measures aircraft, and so on. The types and numbers vary, depending on the size and geographical location of any given air arm. But whatever the country, whatever the air force, the 'unsung hero' is almost inevitably the rotary-winged aircraft . . . the helicopter.

The helicopter has been around for a long time, certainly nowhere near as long as the more conventional fixed-wing types, but with a shorter ancestry which can be traced back to the World War Two era. Although many air forces were quick to seize an opportunity to develop helicopter technology, it was not until the long, bloody and expensive conflict in South-East Asia that the helicopter finally began to be taken seriously as a vital military warplane. In Vietnam the helicopter was used for a variety of tasks, ranging from a flying 'gunship', bristling with machine-guns, to more mundane cargo and equipment transport. But it was the helicopter's vital role as a speedy casualty evacuation aircraft that brought home the serious purpose of the modern helicopter. Countless American servicemen shared the misfortune of being wounded in action during the Vietnam War, but many owe their lives to the helicopter, which plucked them from danger, and delivered them to safety.

Since the 1960s the helicopter has been rapidly developed into a multi-purpose machine, utilized by every military air arm around the world. Indeed, some minor air forces operate nothing but helicopters, such is their versatility. Lightweight and heavy transport is one role that the helicopter performs exceptionally

well, being able to position supplies into areas which might take fixed-wing aircraft (and accompanying vehicles) a vast amount of time to supply. Search and rescue is another vital role, ranging over the world's seas, locating and rescuing hapless victims of sea disasters or abandoned aircraft. Anti-submarine warfare is a more offensive role, dipping sonar listening devices into the oceans, hovering and listening for tell-tale sounds of enemy submarines. Once located, the helicopter launches a devastating attack, using either on-board weapons, or by calling in other friendly forces.

On a similar theme, the helicopter can even become a minesweeper, towing a detector, to clear a sea lane. And high above the sea surface, the helicopter can also become an airborne listening post, fitted with over-the-horizon radar, watching and listening for enemy aerial activity. Vietnam saw the widespread use of helicopters as offensive aircraft, and twenty years later the helicopter gunship is rapidly becoming a 'standard purchase' for many of the world's air forces. Equally, the anti-tank helicopter is quickly being developed as more and more high-technology 'tank-busting' weapons are produced, which can be launched from helicopters.

If Vietnam witnessed the introduction of large-scale rotary-winged military operations, 1991 saw the helicopter become an indispensable part of offensive airpower. Hundreds of helicopters from a variety of nations were deployed to the Middle East during Operation 'Desert Storm', the Allied liberation of occupied Kuwait. Every conceivable kind of helicopter operation was undertaken, and despite early worries and rumours that the 'complicated' helicopter would not be able to cope with desert conditions, reality proved the doubters wrong; and as with many of the new weapons systems that were tried for the first time in 1991, the helicopter emerged as a vital and incredibly successful part of the Allied aerial offensive.

In this book, I have attempted to gather together a wide selection of photographs which show the modern helicopter in various shapes and forms, as seen around the world in the 1990s. Intended to complement *Soviet Military Helicopters* (also published by Arms & Armour Press), this work illustrates every main helicopter type currently in use with the military air arms of the free world. Naturally it would be almost impossible to illustrate and describe every helicopter sub-type based on each basic model, but the most (numerically) important helicopters have been mentioned.

Photographs for this publication were provided by a variety of contributors, all of whom have been credited appropriately. To all who offered their photographs and their assistance, may I offer my thanks, and I hope that the following pages will be a fitting tribute to the ubiquitous helicopter.

3. Being rolled out from the servicing hangar into the heat of the morning sunshine, this Royal Air Force Puma HC.1 is on detachment to Belize International Airport, in support of Army units based there. The Puma has performed well in the stifling heat of Belize. The type is also regularly deployed to Norway for winter training with the British Army, where equally impressive reliability is maintained. The pilots say it is due to the Puma's twin engines, which trace their ancestry to a railway locomotive powerplant! [*J. Webber*]

4. The Aérospatiale Ecureuil (Squirrel) was designed as a direct replacement for the venerable Alouette, but unlike its predecessor, the Ecureuil has yet to enjoy the same amount of sales success. Marketed in the USA as the Astar, this six-seat utility helicopter is operated by French Forces and by Australia, together with a handful of smaller air arms. The type continues to be developed, with continual improvements to payload capability. [*Aérospatiale*]

5. This Squirrel can bite! F-ZKCZ is a late development model of the Ecureuil, referred to as the AS.550 Fennec. Wearing a warlike black and grey camouflage, the aircraft is fitted with the HELITOW anti-tank weapon system. Other weaponry options include 20mm cannon or 68mm rocket pods. The original Ecureuil prototype first flew in June 1974, and as of 1991 the production line is still open. The AS.351 reverts to Aérospatiale tradition by being fitted with a 'fenestron', or enclosed tail rotor. [*Aérospatiale*]

4

5

6. Pictured inside one of RAF Shawbury's servicing hangars, XX382 is one of No 2 Flying Training School's fleet of Gazelle helicopters. The Gazelle is used by the RAF as a basic helicopter trainer, from which students progress to the larger Wessex, before joining a helicopter Operational Conversion Unit, flying one of four RAF front-line helicopter types (Wessex, Puma, Sea King and Chinook). The fleet of British Army and RAF Gazelles was produced by Westland, as part of an Anglo-French co-production deal. [*John Hale*]

7. Demonstrating one of the virtues of Aérospatiale's 'fenestron' enclosed tail rotor design, a Gazelle settles into a patch of woodland camouflage, in perfect anti-tank style. Built with transmission taken from the Alouette, and Bölkow's rigid main rotor (developed for the BO 105), the Gazelle is a successful anti-tank and utility helicopter, in service with countries such as Egypt, Jordan, Libya, Syria and many others, in addition to the main French and British military customers. [*CEV/Aérospatiale*]

8. Wearing French Army Armée de Terre markings, this SA.342L1 Gazelle features an uprated Turbomeca Astazou turboshaft engine, and an improved tail rotor design. As is evident, the Gazelle can carry a variety of armament for various battlefield roles, the available arsenal including two 7.62mm machine-guns, one 20mm cannon, six HOT missiles, rocket pods and wire-guided anti-tank missiles. In this illustration, the Gazelle is fitted with four Mistral air-to-air missiles. [*Aérospatiale*]

9. The Aérospatiale Dauphin (Dolphin) is a replacement design for the Alouette II, and following a first flight in 1972, the type is now in service with France, Saudi Arabia, Eire, and Sri Lanka, among others. The type is also known as the Panther, such an example being this demonstrator model, test firing an Aérospatiale AS.15 anti-ship missile. Used primarily as an anti-ship and SAR (Search and Rescue) helicopter, production of the Dauphin continues. [*Aérospatiale*]

10

11

10. An Aérospatiale AS.565 Panther, in full French Navy (Aéronavale) markings, on a training mission over the French coast. The type is operated from the French aircraft-carriers, primarily on SAR duties, hence the rescue winch attached to the fuselage above the fuselage access hatch. Despite its modest size, the Panther is capable of carrying twelve passengers. [Aérospatiale]

11. Seen at low level, running in on a representative anti-tank attack profile, the Aérospatiale AS.565 Panther is a twin-engined multi-role helicopter in the 4-ton class, fitted with two 20mm pod-mounted cannon for fire support missions. The speckled effect in this photograph is caused by grass cuttings, thrown into the air by the rotor blade downwash. How low can you get! [Aérospatiale]

12. The old but reliable Alouette, still in widespread service, albeit in small numbers, around the world. Main users are the forces of France, Germany and Belgium. The SA.315 Lama, a derivative of the Alouette, is in use with various countries such as Argentina, Bolivia, Chile, India, Morocco and Peru. These two Alouette IIs are French Army examples, carrying AS.11 wire-guided anti-tank missiles. [Aérospatiale]

12

13

13. The Alouette III was the ultimate version of the ubiquitous French helicopter design. With a more powerful engine, the SA.316 and SA.319 are developments of the earlier Alouette II, with a new aerodynamic fuselage shape. Used for a variety of purposes by a staggering number of countries (over 40), the Alouette III is a general-purpose utility/anti-tank/anti-ship and SAR helicopter. 'ABG' is a French Army example, one of 70 such machines. [*Aérospatiale*]

14. Four Aérospatiale Alouette IIIs wearing the distinctive red, white, blue and orange (centre) markings of the Royal Netherlands Air Force helicopter display team, 'The Grasshoppers'. Well known throughout Europe, the team makes regular visits to the UK during the air display season. The R Netherlands AF has 66 such machines, used for liaison duties, and as airborne observation posts. The familiar shape of the Alouette is likely to be around for many years. [*Tim Laming*]

14

15. Out over the Bay of Biscay, an Aérospatiale SA.321 Super Frelon, wearing the distinctive grey and white colours of the French Navy (Aéronavale). The Frelon is a medium-capability transport and anti-submarine (ASW) helicopter, operated by the forces of France, China, Iraq, Israel, Libya, South Africa and Zaïre. Drawing from Sikorsky experience (the rotor system being designed by that company), the Frelon first flew in 1962, and many examples are still in regular service. Aircraft No '60 is an SA.321G Super Frelon fitted with search radar (visible above the main landing gear). [*Aérospatiale*]

16. Another French Navy Super Frelon, one of many now fitted with a large search radar inside a nose-mounted blister. The current low-visibility grey scheme has been compromised on this example by the addition of dayglo orange patches, presumably indicating that this is a training machine. The French Navy operates sixteen Super Frelons, primarily on transport and Search and Rescue duties, from carriers and shore bases. A total of 99 Frelons was built, the last being completed in 1983. [*Kai Anders*]

17. The Agusta A 109 has been a success for its Italian manufacturer, primarily in the civil aircraft field, although military sales continue, albeit on a small scale. The United Kindom became an operator of the type almost by accident, following the capture of Argentine examples during the 1982 Falklands conflict. Four A 109s now serve with the British Army. A general-purpose and light anti-tank helicopter is the A 109EOA, 28 of which are currently operated by the Italian Army. Armament can include eight Hughes BGM-71A TOW anti-tank missiles. 'GN-7947' is one of ten aircraft operated by Venezuela. [*Kai Anders*]

18. Close-up of the Agusta A 129 Mangusta (Mongoose), showing the twin cockpit layout common to many modern anti-tank assault helicopters. The pilot sits to the rear, with the co-pilot/gunner in the forward cockpit, from where he can direct firepower which, in the case of the Mangusta, comprises eight TOW missiles, a 20mm gun pod, and seven 70mm rockets, among a variety of weapons options. First flight was made in 1983, and in addition to 66 machines operated by Italy, aircraft of this type have been delivered to Abu Dhabi. [*Glenn Ashley*]

19. The Agusta-Bell AB.212 is a twin-engined derivative of the AB.205, Italian equivalent of the famous Bell UH-1 'Huey'. The AB.212 also features an improved transmission, structure and systems, and military and civil sales have been many and varied. Built under licence from Bell, the aircraft was developed into the AB.212ASW, a specialized anti-submarine helicopter equipped with a search radar and sonar de-

vices. Customers include Iran, Iraq, Saudi Arabia, Libya and Turkey. Wearing a South-East Asia-style green and brown camouflage, N32113 is a Bell demonstrator. [*Bell*]

20. As the AB.212 was a development of the AB.205, the Agusta-Bell AB.412 is another development, retaining the twin turbine, but with the addition of a four-blade main rotor, which gives a more efficient performance over the earlier two-blade design. The general-purpose military version is known as the Griffon, capable of undertaking general transport duties, casualty evacuation, search and rescue, and tactical support. It can be fitted with a wide range of armament, such as 25mm Oerlikon cannon, TOW anti-tank missiles, or BAe Sea Skua missiles. Customers include Italy and Spain. [*Bell*]

21. Perhaps the most famous helicopter of them all, the Bell UH-1 Iroquois is better known as the 'Huey', thanks to a nickname derived from its initial service designation (later changed) of HU-1A. Aircraft No '53' belongs to the Swedish Army, resplendent in that air arm's unique and rather bizarre splinter camouflage, which even extends to the rotor blades. First flown in 1956, the UH-1 is a multi-purpose assault support helicopter, still in widespread use around the world, and famous for years of excellent service with American units in the Vietnam War. [*Kai Anders*]

22. Later developments of the UH-1 Huey include the UH-1H, operated in small numbers by Argentina. G-HUEY is something of a celebrity in the UK, having appeared at many air shows in civilian hands. The aircraft was captured from Argentina during the 1982 Falklands War, and following refurbishment took to the air once more in support of money-raising efforts for the RAF Benevolent Fund. Sadly the original white colour scheme (yellow band) was obliterated by a sponsor's ugly commercial paintwork. [*Tim Laming*]

23. The Agusta-Bell 206 Jet Ranger is one of the most popular small civil helicopters currently in use. In military guise, the same basic airframe is continuing to be developed into a series of 206 derivatives which have been sold to some 40 countries around the world. Illustrated is one of a small number of 206s used by Morocco for liaison duties, wearing a two-tone brown camouflage and black code numbers. [*Kai Anders*]

24. A beautiful shot taken by one of Bell's photographers of a United States Navy TH-57A SeaRanger cruising over a seascape. The TH-57 is a primary helicopter trainer used by the US Navy since 1968, largely at NAS Whiting Field in Florida. Later improved versions (TH-57B and TH-57C) have now been delivered to the USN. Colour scheme is red and white, with black codes. [*Bell*]

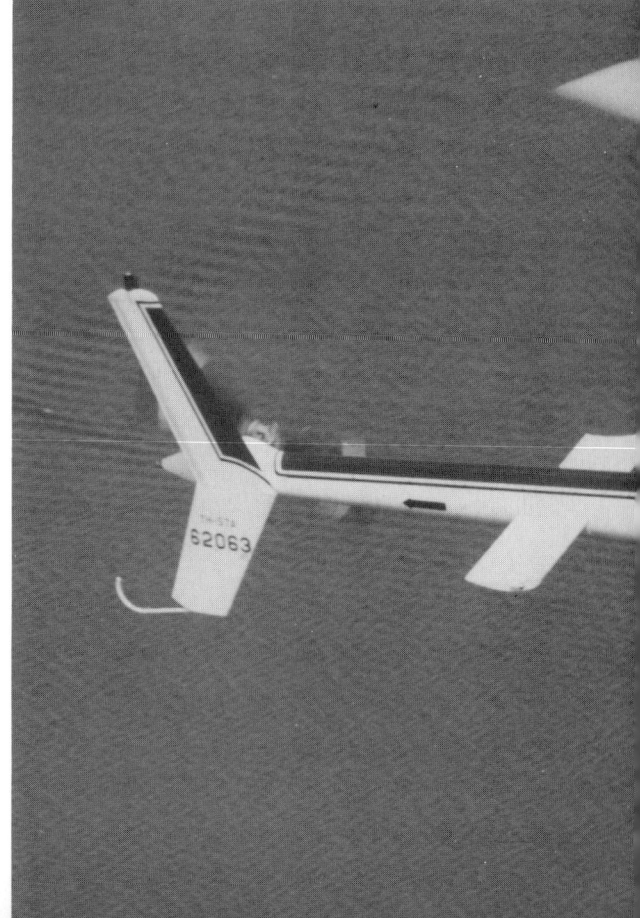

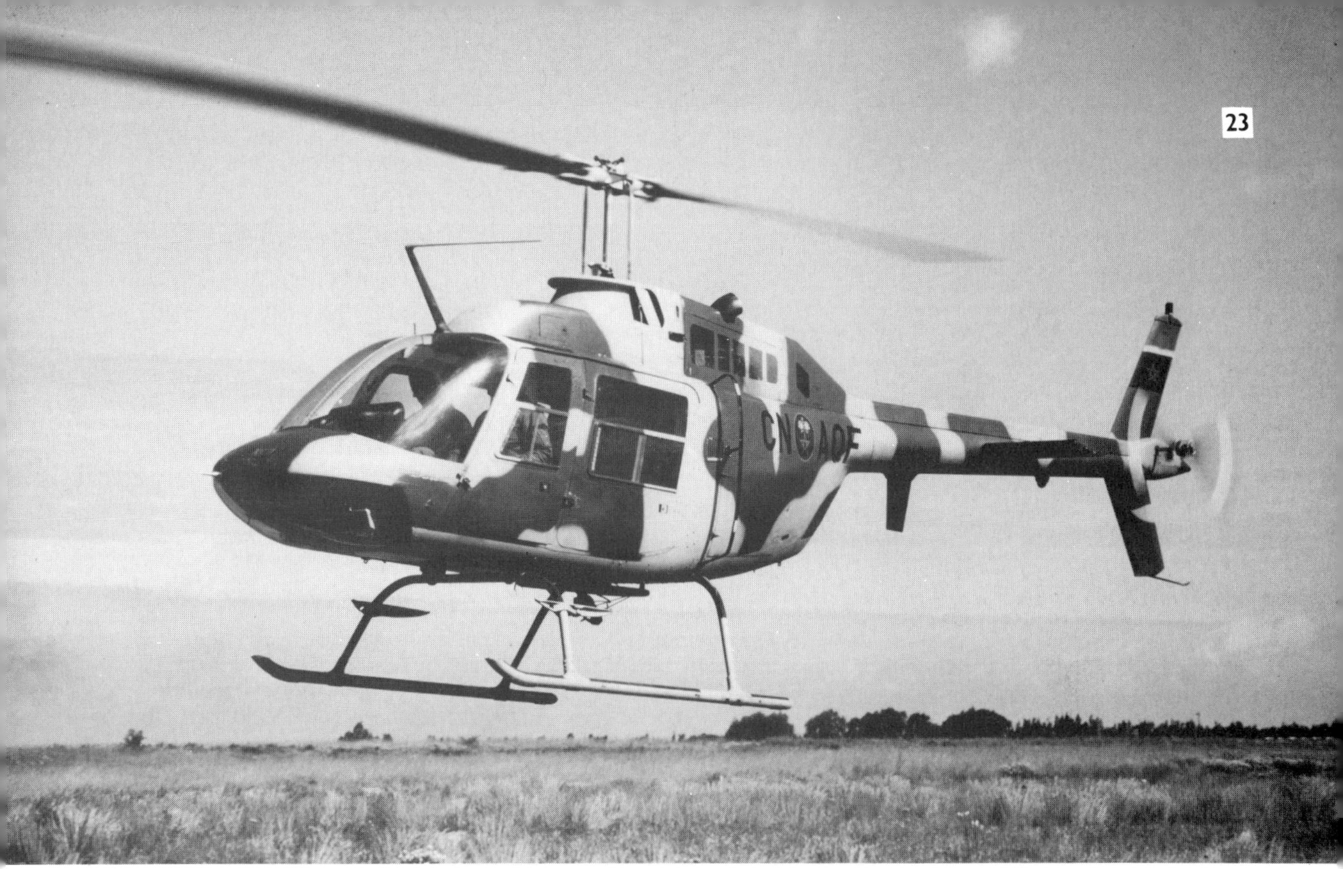

23

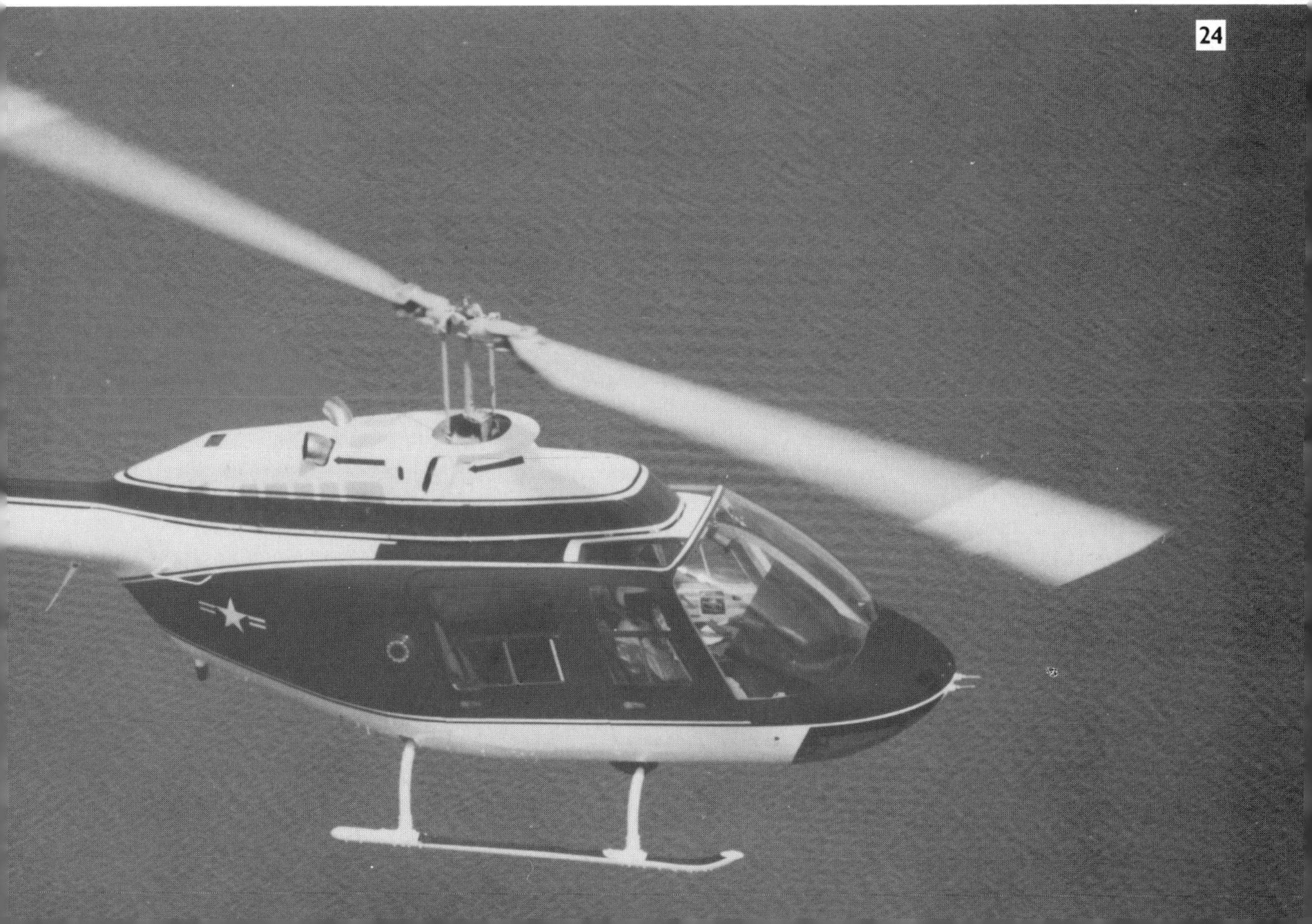
24

25. The Bell 206 is also used in large numbers by the United States Army, as a light observation helicopter. However, it can also be used for casualty evacuation, light transport, photo-reconnaissance, and a variety of utility duties. The basic airframe, designated OH-58A by the Army, is now being deve- loped into the OH-58D as part of an Army improvement programme, giving the helicopter the ability to direct artillery fire and support other attack helicopters. Illustrated is an Army OH-58D complete with rocket packs and a mast-mounted sight. [*Bell*]

26. Saudi Arabia has purchased fifteen Bell 406 Combat Scouts, one such example being photographed at Bell's Arlington facility prior to delivery. Resplendent in two-tone brown camouflage, with green Saudi national markings, the new four-blade main rotor is visible. Considering that the first Bell 206 made its maiden flight in 1966, it is interesting to see that the aircraft is still very much in business 25 years later. [*Bell*]

27. The aggressive-looking Bell HueyCobra became a familiar sight during the 1991 Gulf War, with large numbers of desert-camouflaged machines being seen almost on a daily basis supporting ground forces in their efforts to liberate occupied Kuwait. The AH-1S (illustrated) is an upgraded version of the original AH-1 which first flew in 1965. The 'S-model' features an uprated engine, and is TOW missile-capable. Complete with Army green camouflage, the Cobra is lurking in a woodland clearing. [Bell]

28. As the black titles on the fuselage indicate, this is a US Marine Corps Huey-Cobra, wearing typical Marines light grey, green and black disruptive camouflage. An AH-1W, it is a development of the twin-engined SeaCobra, powered by General Electric T700-GE-401 turboshafts. An impressive array of weaponry can be carried, and the fuel system is designed to withstand 23mm shells. [Bell]

29

29. Out in the desert AH-1W HueyCobras in action. Note that the foremost helicopter is wearing a canopy cover, designed not only to keep out unwelcome sand, but also to keep the internal cockpit temperature down to a reasonable level. Note also the main rotor tie-down cables, attaching the rotor to the nose fairing, above the TOW sighting unit. [*Bell*]

30. Development of the highly successful Cobra continues, the helicopter illustrated being a '4BW', Bell's designation for an AH-1W fitted with a new four-blade main rotor. As can be seen, the new rotor gives the Cobra a very impressive performance, its aerobatic capability being graphically demonstrated. The future of the Cobra is likely to lead to Sidewinder armament and a retractable undercarriage. [*Bell*]

31. Flying for the first time in 1958, the Boeing Vertol Sea Knight is certainly a well established design, but over thirty years later the unusual tandem-rotor helicopter is still in regular service, in large numbers. The US Navy operates over 40 examples of the HH-46A (illustrated performing a practice rescue mission), while the US Marines currently maintain a fleet of over 300 CH-46Es, primarily as transport helicopters. [*Boeing Vertol*]

30

32. The Swedish Navy operates ten KV-107s on anti-submarine warfare and transport duties, one such aircraft being illustrated, wearing the usual eye-catching splinter camouflage favoured by Swedish forces. Also visible are white 'Marinen' titles, a yellow tail code, and dayglo orange patches. The Swedish Air Force also flies nine KV-107s on search and rescue missions. [*Kai Anders*]

33. The impressive sight of the Boeing Vertol Chinook is matched only by the equally distinctive 'whump, whump' of its huge rotor blades. This example of No 7 Squadron, RAF, demonstrates the Chinook's airlift capability, heaving Army AFVs into the air. Capable of carrying 16,000lb externally, the Chinook is currently in service with the forces of Canada, Japan, Spain, South Korea, United States, United Kingdom and Vietnam, among others. [*Tim Laming*]

34. An RAF Chinook with forward access door and rear cargo door open. It can carry 44 troops. The RAF deployed pink-camouflaged Chinooks to the Middle East during the 1991 Gulf War; 36 Chinooks are at present in service with the RAF, in support of the British Army. The US Army operates over 550 'C' and 'D' models. [*John Hale*]

35. Still in the mock-up stage, this Eurocopter Tiger illustrates the shape of the forthcoming Franco-German anti-tank helicopter design. Designed and built by Messerschmitt-Bölkow-Blohm and Aérospatiale, the Tiger will be built in three basic versions. One will be a fire support and escort machine for the French Army, the second an anti-tank version for the German Army, and the third another anti-tank variant destined for service with the French Army. [*MBB*]

36. The curious shape of the Kaman H-2 Seasprite has been around for a long time, the first aircraft taking to the air for the first time in 1959. Still very much in service 30 years later, the Seasprite is the US Navy's lightweight multi-purpose helicopter, used primarily for anti-submarine warfare and search and rescue duties, operating from warships and land bases. The UH-2C illustrated is no longer in service, the remaining 50 examples being SH-2Fs, many now repainted in tactical grey camouflage. [*via Tim Laming*]

37. The Kaman H-43 Huskie was once a familiar sight, in widespread service with the USAF during the Vietnam War, as a search and rescue helicopter. Long since retired from USAF service, small numbers of Huskies continue to serve the forces of Pakistan (four machines) and Burma (seven). The Huskie was a versatile helicopter, able to carry ten passengers, and could undertake ambulance, firefighting and liaison duties as well as its primary SAR task. [*USAF*]

A. Grey Whale – a Westland Sea King AEW.2 banking at low level over RAF Mildenhall in May 1989. The low-visibility grey camouflage and red/blue national insignia, plus black titles, were introduced on all front-line Fleet Air Arm aircraft following the 1982 Falklands War. Sadly the Sea King AEW.2 was not developed until after the conflict, and with hindsight (on the part of the Government, not the RN) it was realized that the Navy badly needed airborne radar cover. [*Tim Laming*]

B. Running-in low over RAF St. Mawgan's runway in August 1989, a Chinook HC.1 from the Operational Conversion Unit based at RAF Odiham. The RAF currently has 36 Chinooks used for heavy transport duties, in support of the British Army. A small number of machines drawn from Nos 7 and 18 Squadrons (based at Odiham and Gütersloh respectively) were sent to the Middle East as part of Operation 'Granby', the RAF contribution to the Allied Operation 'Desert Storm', in 1991. [*Tim Laming*]

C. Wearing celebratory markings for HFR 35's anniversary, a West German Sikorsky CH-53G. Two 'G' models were built for West Germany by Sikorsky, with a further 110 being assembled by VFW-Fokker. The type is basically the same as the CH-53D, with capacity for 55 troops, a maximum speed of 170 knots, and a range of 275 miles. [*Christian Gerard*]

A▲

B▲　C▼

D. Wearing Italian military markings, the Agusta A 109 Hirundo is one of 28 aircraft currently in Italian service. Two aircraft are being used for trials purposes, fitted with various weapons systems such as the TOW missile, together with a chin-mounted sight. The Agusta A 109 can also be fitted with a 7.62mm machine-gun, 68mm rockets, and various other armament fits. [*Giuseppe Fassari*]

E. The Royal Navy's Sea King HAR.5 wearing light grey camouflage but with the addition of bright red high-visibility colours for its search and rescue task. XV705 '821' is one of four such Sea Kings currently operating with No 771 Naval Air Squadron at RNAS Culdrose, Cornwall. It is pictured in a hover over RAF St. Mawgan's runway in Au-

▲ **D**

gust 1990, while performing a public demonstration of a casualty rescue by winch. [*Tim Laming*]

F. Sea King ZB506 is a rare example of the Royal Aerospace Establishment's Sea King 4X, a specialized derivative based on the

Commando-configured Mk. 4, but with the addition of the Mk.5's search radar. This particular machine is operated by the RAE at Thur-

▼ **E**

leigh near Bedford, and when photographed in July 1989, the aircraft was fitted with a trial radar system in a rather odd-looking forward housing attached to the nose. Six Sea Kings are currently operated by the RAE at Bedford, Farnborough and Boscombe Down. [*Tim Laming*]

F ▲

G. A line-up of Connecticut Army National Guard CH-54 Skycranes at Bradley Airport, their home base, shortly before being replaced by Chinooks. Maximum take-off weight for this peculiar outsize machine is 42,000lb. The main rotor diameter is 72ft and the overall length (including rotors) is 88ft 6in. The externally mounted engines are Pratt & Whitney T73-P-1 turboshafts, each providing 4,500shp. In addition to the small number of Skycranes still used by the Army National Guard, a few examples are operated around the world by specialized civil firms. [*Tim Laming*]

G ▲   H ▼

H. A Sikorsky HH-3E belonging to the USAF's 56th ARRS, based at Keflavik in Iceland. The squadron provides search and rescue and combat rescue cover for most of the USAF's operations in Europe. The HH-3 is a rescue version of the SH-3 designed for the US Navy, which was later purchased by the USAF in an advanced version, designated CH-3, the first being delivered in 1963. The HH-3 was a later, improved version which served in Vietnam, where it gained the familar nickname 'Jolly Green Giant'. The US Coast Guard purchased an unarmed, unarmoured version designated HH-3F Pelican, which was later also purchased by the Italian Air Force for SAR duties. [*Scott Van Aken*]

▲I ▼J

I. A Bell UH-1N belonging to El Toro Marine Corps Air Station, California, returning from a training mission over the southern California coast. Powered by a Pratt & Whitney Canada T400-CP-400 rated at 1,290shp, maximum speed is 110 knots, and maximum take-off weight 10,000lb. The engine used in the UH-1N is essentially two identical engines, mounted side-by-side and connected to a single driveshaft by a series of gears. [*Tim Laming*]

J. The Swedish Army operated a fleet of nineteen Agusta-Bell AB.206A Jet-Rangers on utility duties. Like almost every other Swedish military aircraft type, the JetRanger has been camouflaged in the usual splinter colour scheme, comprising two tones of green, brown and black. Maximum speed is 114 knots, range 527 miles. Maximum take-off weight is 3,350lb. No armament is car-

K▲

ried, the type being essentially the same as the well known civil version [*Kai Anders*]

K. A pair of Sikorsky designs together, both licence-built and developed by Westland in England. At the rear is a Wessex HC.2 from the RAF's No 22 Squadron, while at the front is a Sea King 4X, ZB507, belonging to the Royal Aerospace Establishment at Farnborough. The Wessex is a search and rescue machine, while the Sea King serves as an experimental airframe on a variety of test and transport duties.

Both were used during the 1988 SBAC show at Farnborough as fire-fighters. [*Tim Laming*]

L. A rare illustration of an Aérospatiale SA.321 Super Frelon of the Libyan Air Force. Twelve such aircraft are used by Libya for anti-submarine warfare and search and rescue. Note the addition of engine intake sand filters, and the absence of search radar found on French machines. This particular aircraft was pictured during a visit to RAF Luqa, Malta, during the early 1970s. [*Richard J. Caruana*]

L▼

▲M

▲N ▼O

M. Resplendent in the striking dayglo orange and green colours of the French Army (Armée de Terre), an Aérospatiale SA.342M Gazelle. This is an advanced version of the Gazelle developed for the anti-tank role. It is fitted with the Astazou XIVM turboshaft (with automatic start), and a revised instrument panel layout as specified by ALAT (Aviation Légère de l'Armée de Terre). Also included is an autopilot, doppler radar and a gyro-stabilized sight for the guidance of HOT missiles. A 20mm cannon or two 7.62mm machine-guns can also be carried, or a pair of AS.12 missiles. [*Christian Gerard*]

N. On the shores of the Pacific, an Aérospatiale SA.366/HH-65A Dolphin at its home base of San Diego US Coast Guard station. Although essentially the same airframe as the SA.365N, the HH-65 has two Lycoming LTS101-750A-1 turboshaft engines, and other US-manufactured equipment (up to some 60 per cent of the total aircraft cost). Equipped for a three-man crew, plus passengers, the Dolphin is fitted with flotation bags which enable the USCG crews to land on the sea even in fairly rough conditions, safely to evacuate the aircraft in an emergency. The helicopter is fitted with a FLIR (Forward-Looking Infra-Red) detector for night SAR operations. [*Tim Laming*]

O. Unusual by any standards, the Schweizer (Hughes) 300C is operated by the Swedish Army, primarily on training duties. The type also serves with the forces of Algeria, Argentina, Brazil, Colombia, Greece, Iraq, Japan, Thailand and Turkey, among others. A small, simple and

inexpensive helicopter, the 300C is used for various utility and training duties. The US Army was an operator of the type (TH-55) until the UH-1 was introduced as the Army's basic helicopter trainer in 1988. [*Kai Anders*]

P. MM80999 is one of fifty McDonnell Douglas NH-500Ds currently in service with the Italian Air Force. A successful civil lightweight helicopter, the MDD (Hughes) 500 has, in a wide variety of versions, been purchased by approximately 20 different military air arms, with sales of the type still continuing. With a length of just over 30ft, maximum cruising speed is 119 knots, the powerplant being a single 420-shp Allison 250-C20B turboshaft. [*Luca Storti*]

Q. Fly Navy! The Fleet Air Arm's No 705 Naval Air Squadron is the Royal Navy's Gazelle unit, based at RNAS Culdrose in Cornwall. The squadron is responsible for the initial and basic helicopter flying training of all FAA pilots, and over 20 examples are normally on strength with the unit. No 705 NAS also provides the European air show scene with a helicopter display team, known as 'The Sharks'. One of their Gazelles is illustrated, performing some low-level aerobatics during the 1988 Mildenhall Air Fete. [*Tim Laming*]

▲R

▲S ▼T

R. A CH-46D Sea Knight of the US Navy's HC-11 awaiting engine start-up clearance at Miramar Naval Air Station, California, in August 1990. The US Navy operates over 40 Sea Knights on both transport (as illustrated) and search and rescue duties. Unlike their USMC counterparts, who apply disruptive camouflage to their Sea Knights, the US Navy machines wear an overall dark-grey gloss polyurethane scheme. [*Fred Mugford*]

S. Switzerland maintains some 21 Alouette IIs for liaison duties, and another eighteen military air arms also continue to fly the SE.3130/SA.318C on various light transport and other general-purpose tasks. Powered by a Turbomeca Astazou IIA turboshaft, maximum speed is 110 knots, with a service ceiling of 10,827ft, a figure which is worth considering, particularly in the Swiss mountains. Note the Alouette III in the hangar. [*Paul Hoehn*]

T. Looking mean in its tactical medium-grey camouflage, a Sikorsky SH-60B Seahawk, one of 204 'Bravo' models for the US Navy. Maximum speed is in excess of 125 knots, thanks to the power supplied by two General Electric T700-GE-401s, each developing 1,690shp. Pictured on the arrivals ramp at Pease AFB in New Hampshire during May 1990, the Seahawk is a familiar sight across the USA, and the type played a vital role in the 1991 Operation 'Desert Storm', flying anti-submarine missions, search and rescue flights, and other transport and liaison duties. [*Joel Paskauskas*]

38. Air combat manoeuvres were demonstrated in public by the AH-64 Apache during the 1988 Farnborough Air Show. The display routine was flown by a McDonnell Douglas test pilot, and included 360deg rolls, a one-half Cuban eight, descending spirals and a hammerhead manoeuvre, the kind of aerobatic exercises normally reserved for fixed-wing aircraft. The aim was to demonstrate the Apache's flexibility for air-to-ground and air-to-air combat missions. [*McDonnell Douglas*]

39. An awesome firepower display laid out in front of the AH-64. The weapons on show include Sidewinders and Sidearm air-to-air missiles, the Mistral AAM, 1,200 rounds of 30mm ammunition, 76 × 2.75in (70mm) rockets, 16 Hellfire anti-tank missiles (on the inboard pylon stations), wingtip-mounted Stinger air-to-air missiles, and external fuel tanks. Not a bad collection of equipment for one combat helicopter! [*McDonnell Douglas*]

40. An Apache equipped with the Longbow integrated fire control radar and missile system, emerging from behind a mountain during highly successful US Army tests at the Yuma Proving Ground in Arizona. Consisting of a millimetre wave radar fire control system mounted on the main rotor mast, and a 'fire and forget' Hellfire missile, Longbow provides all-weather capability and long-range anti-armour targeting ability for US Army helicopters. On the nose is the Target Acquisition Designation Sight/Pilot Night Vision Sensor (TADS/PNVS). [*Martin Marietta*]

41

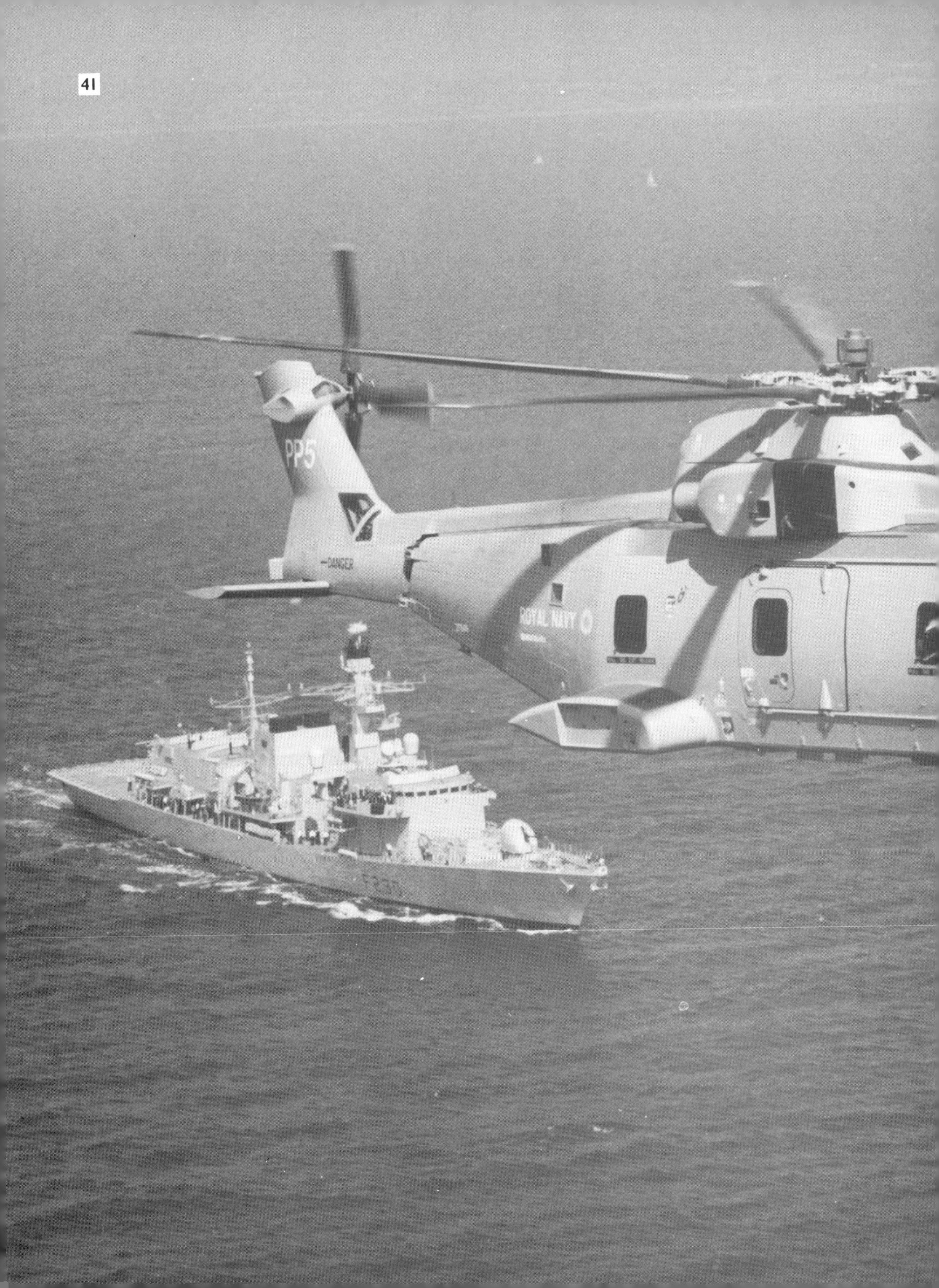

41. The shape of things to come, the EH Industries EH 101, an Anglo-Italian venture which will lead to a new helicopter, destined to replace Royal Navy Sea Kings and RAF Pumas. PP5 is one of the British pre-production models built by Westland. With HMS *Norfolk* in the background, the EH 101 wears what is expected to be the standard Royal Navy light grey colour scheme. The naval helicopters will be called 'Merlin'. [*Westland*]

42. In Royal Air Force service the EH 101, called 'Griffon', will replace Pumas as the main tactical support helicopter for the British Army. In this picture a Westland-built mock-up demonstrates the kind of battlefield scenario in which the Griffon is designed to operate. The EH 101 is a truly international venture, designed and built by Westland and Agusta, with Rolls-Royce/Turbomeca engines. In addition to Royal Navy and RAF machines, the Italian Navy will also take delivery of the aircraft. [*Westland*]

43

44

43. The Hughes 369 illustrated is one of fourteen examples operated by the Danish Army on observation, liaison and reconnaissance duties. The US Army's OH-6 is similar and more than 330 remain in service as light observation machines. The type is also operated by the Dominican Republic, Japan, Nicaragua and El Salvador. One OH-6A was modified by Hughes (now part of McDonnell Douglas) into a NOTARR (NO TAil RotoR) test aircraft, a programme which is still being pursued by the manufacturer. [*Glenn Ashley*]

44. The Philippine Air Force has added the McDonnell Douglas 500MD Defender to its helicopter fleet, joining a growing number of nations now flying this combat-proven aircraft. Twenty-two Defenders were ordered through a foreign military sales agreement between the Philippine and US governments. Wearing two-tone brown and green tactical camouflage, the Philippine national insignia is applied in black outline on the tail boom. [*McDonnell Douglas*]

45. The current NOTARR design is based on the MDD 530N model, a development of the earlier Hughes 500. The tail rotor is replaced by high-pressure air which is bled from the engine and released through small vents in the tail. The vents can be varied in size, thus creating a varying amount of aerodynamic lift over the tail. The 'lift' is directed horizontally, thus maintaining directional stability. The system is now functional and may well become a common feature of future helicopter design. [*McDonnell Douglas*]

46

47

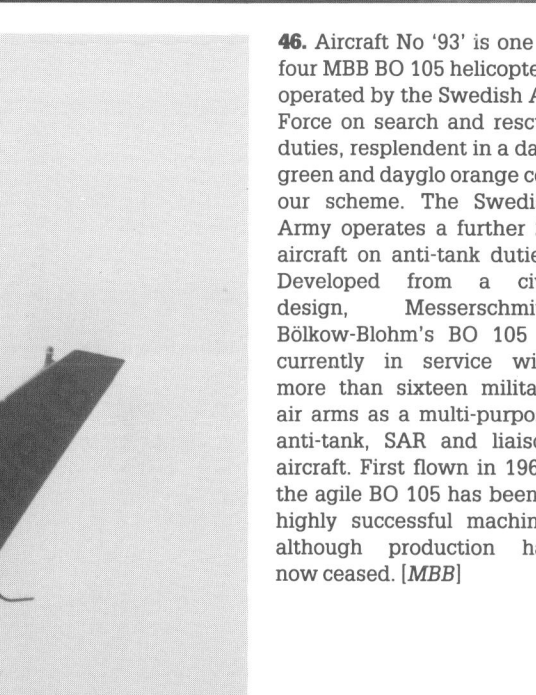

46. Aircraft No '93' is one of four MBB BO 105 helicopters operated by the Swedish Air Force on search and rescue duties, resplendent in a dark green and dayglo orange colour scheme. The Swedish Army operates a further 20 aircraft on anti-tank duties. Developed from a civil design, Messerschmitt-Bölkow-Blohm's BO 105 is currently in service with more than sixteen military air arms as a multi-purpose anti-tank, SAR and liaison aircraft. First flown in 1967, the agile BO 105 has been a highly successful machine, although production has now ceased. [*MBB*]

47. Following experience with the successful BO 105, MBB developed and built, in conjunction with Kawasaki, the BK-117, a civil helicopter (with 11-seat capacity) which was later modified for the military market. Sales so far have been small, although production continues. D-HBKB (illustrated) is an MBB development aircraft fitted with a Lucas gun turret, painted in a 'tactical' green colour scheme. Truly a multi-role design, the BK-117 is sure to enter service with various air arms in the future. [*MBB*]

48. Although it is common practice for manufacturers to produce larger and more powerful versions of earlier aircraft designs, the Sikorsky S-62 took a reverse path, being designed as a scaled-down version of the earlier S-61. In the belief that there was a market for a smaller capacity helicopter of the same type as the S-61, the result was an efficient helicopter for which there was, in reality, little requirement. The United States Coast Guard did take delivery of a modified version (HH-52A, as illustrated), and small numbers may remain in use outside the USA. [*Tim Laming*]

49. Heavylift mobility: two US Marine Corps Sikorsky CH-53E Super Stallion heavylift helicopters taking on fuel from a USMC KC-130 Hercules. Both machines are carrying underslung loads of light armoured vehicles. The CH-53 first flew in 1974, and the 'E' model is capable of carrying up to 32,000lb of external cargo, which includes the M-198 155mm howitzer. Internally there is capacity for 55 fully equipped combat troops. The type was used extensively during the 1991 Gulf War. [*Sikorsky*]

50

51

50. Unusual by any standards, the Sikorsky S-64 Sky-crane is, as its name suggests, literally a flying crane. First flown in 1962, it is designed to carry a variety of extremely heavy and outsize loads, which are underslung externally, rather than having to be fitted inside the confines of a fuselage. As illustrated, the internal accommodation is very small indeed, with room for just a pilot and co-pilot, with a rearward-facing seat to allow careful control of load positioning. Even the engine structure conforms to the 'inside-out' design concept. [*Kai Anders*]

51. Pictured with the Connecticut Army National Guard shortly before replacement by Boeing Chinooks, this CH-54 (military designation for the S-64) carries an underslung pod, which instantly converts the helicopter into a more traditional troop carrier, capable of accommodating 67 troops or 48 stretchers. Alternative loads can include a mobile field hospital, or 22,890lb of cargo. Over 70 examples remain in use exclusively with the US Army, with National Guard units. The type saw extensive service during the Vietnam War and earned a degree of fame by carrying outsize bombs, used for instantly clearing woodland to create helicopter landing sites. [*Tim Laming*]

52. Another derivative of the successful CH-53 design is the MH-53E Sea Dragon. This example is operated by HM-12, an Atlantic Fleet mine-hunting squadron of the US Navy based in Norfolk, Virginia. The Sea Dragon detects enemy mines by towing a detection 'sled' over the sea surface. The Airborne Mine Countermeasures (AMCM) MH-53E is one of the latest developments in a long line of Sikorsky S-65 derivatives, in service with the USN, USMC, USAF, and with Israel, Germany and Iran. [*Sikorsky*]

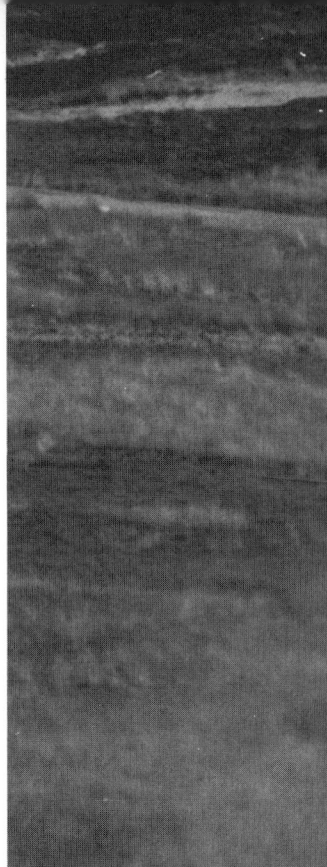

53. Developed in response to the US Army's Utility Tactical Transport Aircraft System (UTTAS) requirement, the Sikorsky S-70 first flew in 1974, being given the service designation UH-60. The design became the familiar Blackhawk, now entering service with the US forces in vast numbers. The US Army anticipates delivery of more than 2,000 examples. Illustrated is a Blackhawk carrying a typical AFV underslung load. [*Westland*]

54. Showing something of the offensive capability of the UH-60, this Blackhawk is low down over a British firing range, releasing a salvo of 76 × 2.75in rockets. Other armament options include Hellfire ASMs, mine-dispensers, jamming flares, chaff-dispensers, and 7.62mm machine-guns, mounted inside the cabin. Initially the Blackhawk suffered from technical problems, but its unserviceability difficulties are being overcome and the type proved to be a successful combat assault transport helicopter, used during the Panama and Middle East conflicts. [*Westland*]

55. Production of the UH-60A Blackhawk continues at Sikorsky's Stratford plant in Connecticut. Capable of carrying a fully equipped 11-man infantry squad, external loads of up to 8,000lb can also be lifted. Deliveries of EH-60s is also being made, these being specialized electronic counter-measures (ECM) and electronic surveillance measures (ESM) machines. The USAF also expects to receive deliveries of the HH-60A Night Hawk combat rescue derivative. [*Sikorsky*]

54

55

56. Strike rescue and special warfare operations are the primary missions of the new US Navy HH-60H Helicopter Combat Support (HCS) aircraft. The basic navalized derivative, the SH-60B Seahawk, first flew in 1979 and is now in widespread service with the USN, performing anti-submarine warfare and search and rescue duties from ships and shore bases. Nearly 400 Seahawk variants are expected to be built for the USN. [*Sikorsky*]

57. Yet another Sikorsky S-70 derivative, the HH-60J Jayhawk is a medium-range recovery helicopter now entering service with the US Coast Guard, replacing the elderly fleet of Pelican helicopters around the USA's coastline. Capable of flying a 300 nautical mile radius from base, the helicopter can recover six passengers and have an on-scene search and rescue time of 45 minutes. Other missions will include drug interdiction and environmental clean-up support. [*Sikorsky*]

58. What may become one of the most famous Blackhawks, this is a VH-60, one of nine similar helicopters operated by the US Marine Corps, on VIP and media transport duties, most notably providing the personal transport for the President, as required around the world. It is likely that this green and white VH-60 will eventually become a familiar sight landing in the grounds of the White House. [*Sikorsky*]

58

59. Essentially a civil helicopter design, the Sikorsky S-76 Spirit has since 1975 been marketed as a potential military general-purpose helicopter. So far sales have been small, to countries such as Jordan (14 aircraft), Philippines (15 aircraft), Brunei and Honduras. The type is used for search and rescue, liaison and other general transport duties. The H-76 Eagle is a specialized military derivative with armour plating, weapons pylons, attack avionics and mast-mounted sight; however, no sales have yet been made of this version. Illustrated is a pre-delivery example of the S-76. [*Sikorsky*]

60. The familiar shapes of two Westland Lynx AH.7s of No 669 Squadron, Army Air Corps, flying low over their 'home ground' in West Germany. The Lynx was one of three helicopters designed and built under an Anglo-French programme between Aérospatiale and Westland (the other two types being the Gazelle and Puma). Initially aimed at naval and civil use, the Lynx was later developed for the Army as a multi-role transport and anti-tank attack helicopter.

Both the machines shown here are armed with TOW missiles and wear what is now the standard AAC grey/ green camouflage, replacing the early black/green colours. [*Westland*]

61. Wearing overall medium grey paintwork, ZE477 was Westland's Lynx 3 demonstrator, a dedicated anti-tank derivative with uprated Rolls-Royce Gem turboshaft engines and an advanced-technology four-blade main rotor, together with a slightly longer fuselage. Capable of firing Hellfire, TOW or HOT missiles, rockets and cannon, the illustration shows a fit of four Rockwell Hellfire missiles. [*Jan van Ommen*]

62. The latest British Army derivative of the Lynx is the AH.9, now entering AAC service. Sixteen have been ordered together with a further eight which will be converted from older AH.1s and AH.7s. Note that the AH.9 has a strong fixed undercarriage, as opposed to the skids of earlier AAC Lynx versions, and the latest advanced main rotor design. Note also the exhaust shrouds and the lack of external armament in this picture, although the type is capable of carrying a wide range of weaponry. [*Westland*]

63. Pictured on the rolling deck of HMS *Alacrity*, the Lynx HAS.3 is the naval derivative used by the Fleet Air Arm for anti-submarine warfare. More than 80 are currently in use, primarily aboard the Royal Navy's Type 21, 22 and 42 class destroyers. Using the negative-6° rotor pitch, locked brakes, and a harpoon deck-lock system (which physically holds the helicopter on to the deck), this Lynx is coping with some pretty rough sea conditions, prior to launching on an anti-ship attack mission. [*Westland*]

64. Still in business, albeit in small numbers (around 30), the Westland Scout AH.1 has served the British Army for many years as a light-weight utility helicopter. First flown in 1958, the Scout was originally designed by Saunders-Roe, before that company was absorbed into Westland. The definitive Scout AH.1 entered AAC service in 1963, and 105 aircraft were produced, serving around the world in support of various Army operations. Despite being largely replaced by the Lynx, the useful little Scout is still in service, and likely to remain so for some time. [*MoD PE*]

65. A 1977-vintage illustration of a Sea King HAS.1 of No 819 Naval Air Squadron on a training flight from its home base of HMS *Gannet* (Prestwick Airport), wearing the original Navy colour scheme (dark grey) with white codes and full colour roundels. The Westland Sea King was a licence-built version of Sikorsky's S-61, developed as a 'hunter-killer' helicopter for the Royal Navy. The type first flew in 1967 and following initial entry into service, Westland have continued to develop the Sea King into a number of variants, giving the original airframe a whole range of new capabilities. [*Royal Navy*]

66. Later in the Sea King's career, the type has now been embarked in some of the Navy's smaller ships and, as illustrated, fitting a relatively large helicopter on to a small deck is quite a problem. The Sea King HAS.1 was developed into the HAS.2, with uprated engines and a six-blade tail rotor. The Sea King HAS.5 illustrated is identified by a new, large dorsal radome (just visible), housing a new search radar. Note also the rescue winch on the starboard fuselage. [*Westland*]

67. The Sea King HAR.3 was developed for the RAF as a search and rescue helicopter, replacing the aged Whirlwind, and is now in regular service alongside the Wessex. Carrying a crew of four, the HAR.3 has capacity for 19 passengers, two stretchers and 11 passengers, or six stretchers. A detachment of RAF Sea Kings remains at RAF Mount Pleasant on the Falkland Islands. Although the rescue yellow colour scheme is common, some aircraft (notably the Falklands machines) wear a tactical grey paint scheme. [*Westland*]

68. Westland's 'Advanced Sea King', fitted with the latest technology available, showing that the basic Sikorsky S-61 airframe is still capable of further improvement. Export versions include the Commando, a tactical transport helicopter with a fixed, strengthened undercarriage. The Sea King AEW.2 carries a huge airborne early warning radar on the starboard fuselage, housed in an inflatable hinged radome. Egypt, Australia, West Germany, Norway, Pakistan, Belgium, India and other countries have all adopted Westland's highly successful Sea King. [*Westland*]

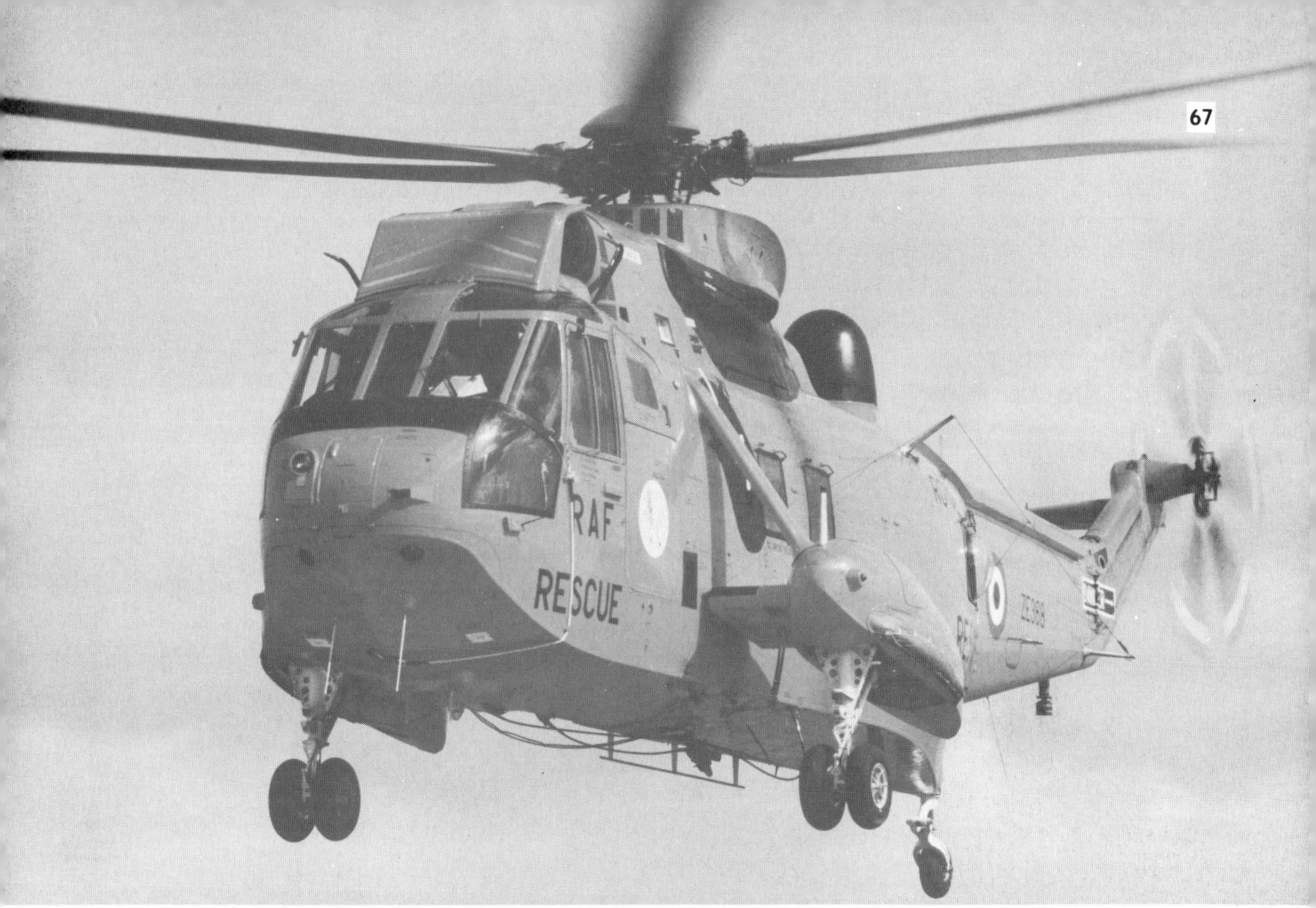

67

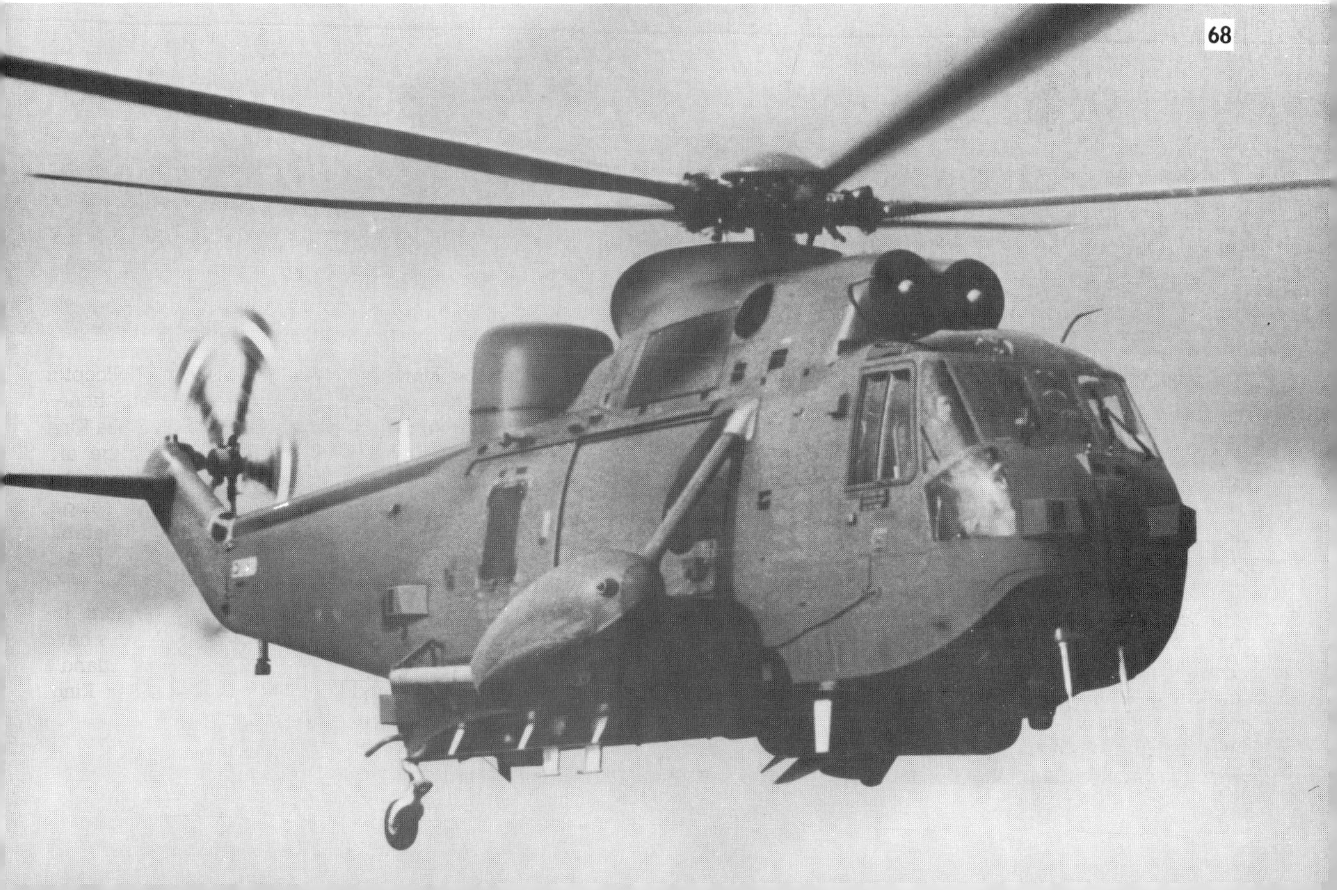
68

69. As the British Army adopted the Westland (Saunders-Roe) Scout, the Royal Navy purchased the Wasp, a navalized version with wheeled landing gear as opposed to skids. Entering service in 1963, 98 aircraft were produced for the Royal Navy, XV362 being one such machine, pictured shortly before retirement from FAA service. The tactical colour scheme of black codes and red/blue roundels was a direct result of experience in the 1982 Falklands conflict, in which a number of Wasps took part. [*Tim Laming*]

70. XV541 wears an early colour scheme, complete with high-visibility dayglo orange patches, often applied to Wasps serving aboard the Antarctic survey ship HMS *Endurance*. Largely withdrawn from service in 1988, a few Wasps remained in use for a further two years, but the type has now been completely retired. However, overseas operators (Brazil, Indonesia and New Zealand) continue to operate the diminutive Wasp for anti-submarine and anti-ship training, search and rescue, and other utility duties. [*Royal Navy*]

71. The Westland Wessex is a familiar sight in the United Kingdom, having served with the Royal Navy and RAF for over 30 years. Developed from the Sikorsky S-58, the Wessex HC.2 and HAS.3 (which remain in use) are powered by Rolls-Royce Gnome turboshaft engines. Although long since withdrawn from RN service, the RAF maintains a fleet of search and rescue helicopters, together with a tactical support squadron based in Northern Ireland. The RAF helicopter training unit No 2 FTS also uses a small number of Wessex for advanced training. [*Tim Laming*]

70
XS 541 ROYAL NAVY
81
DANGER

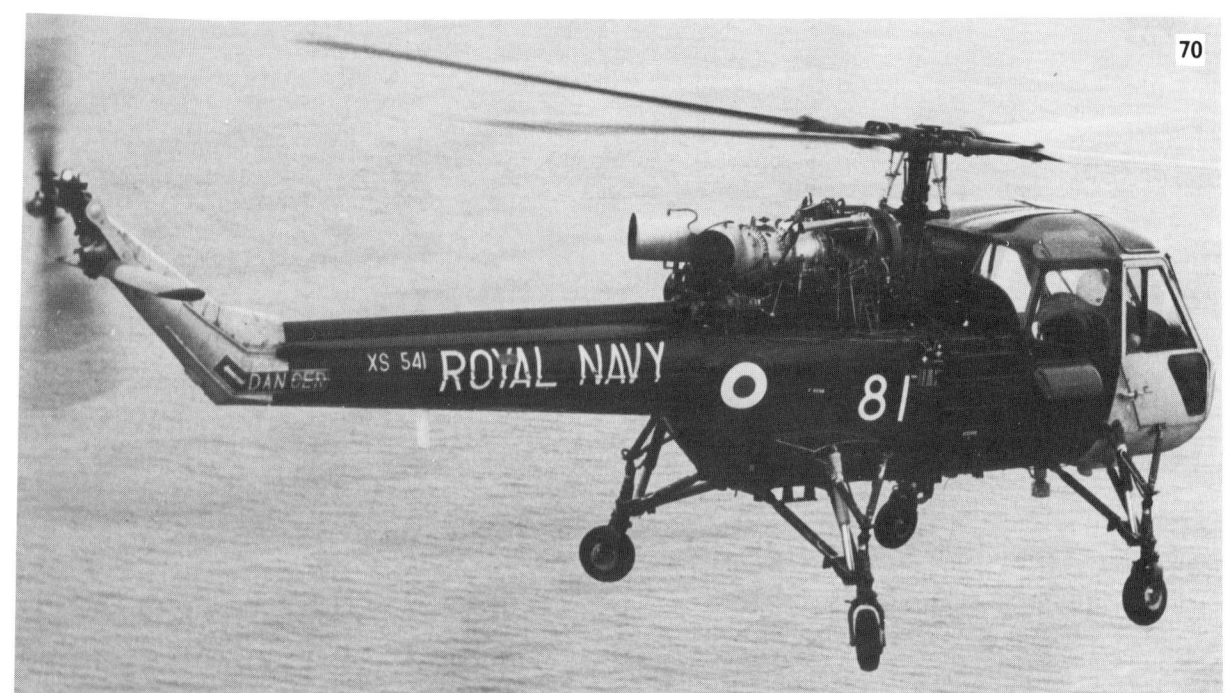

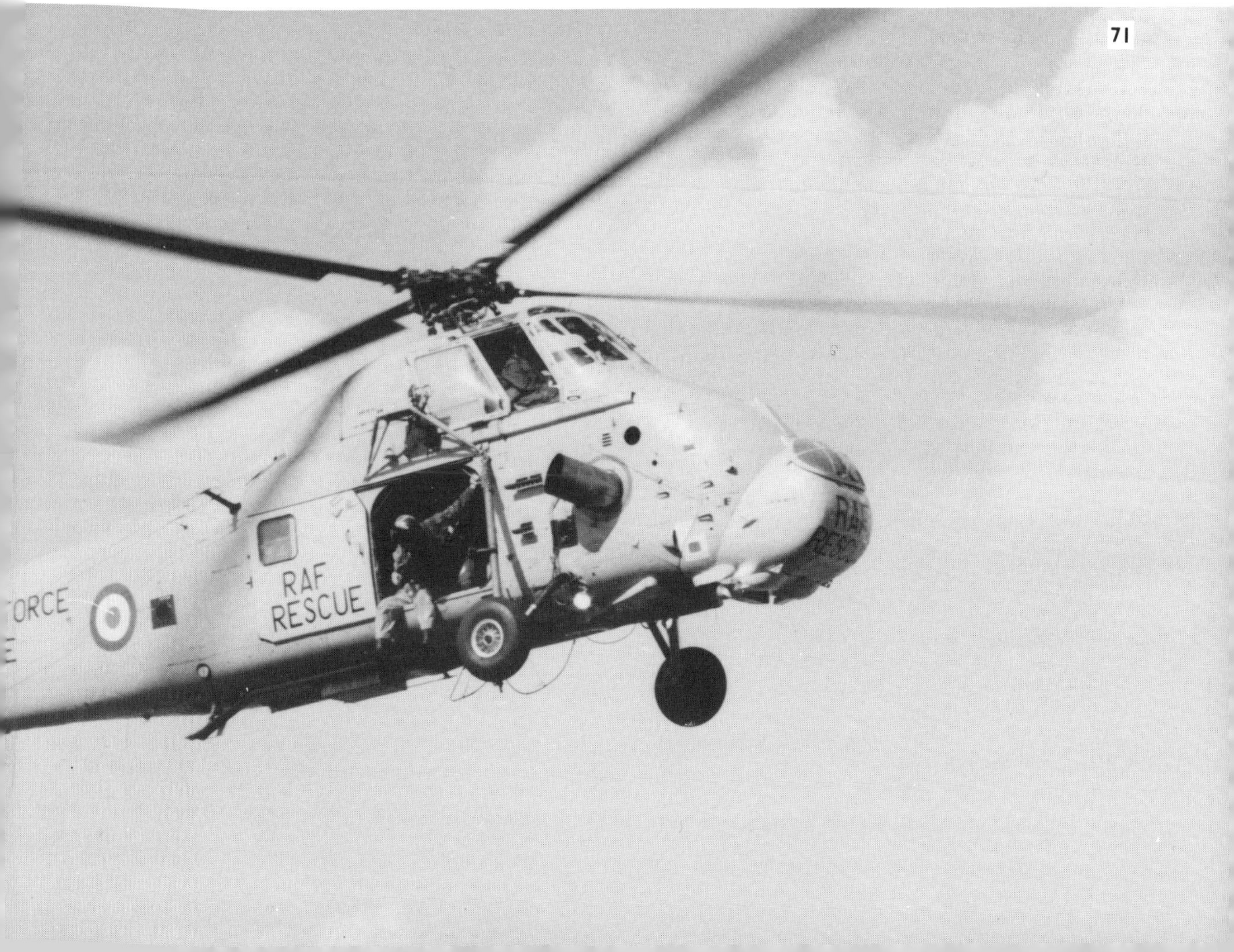
71
RAF
RESCUE
ORCE

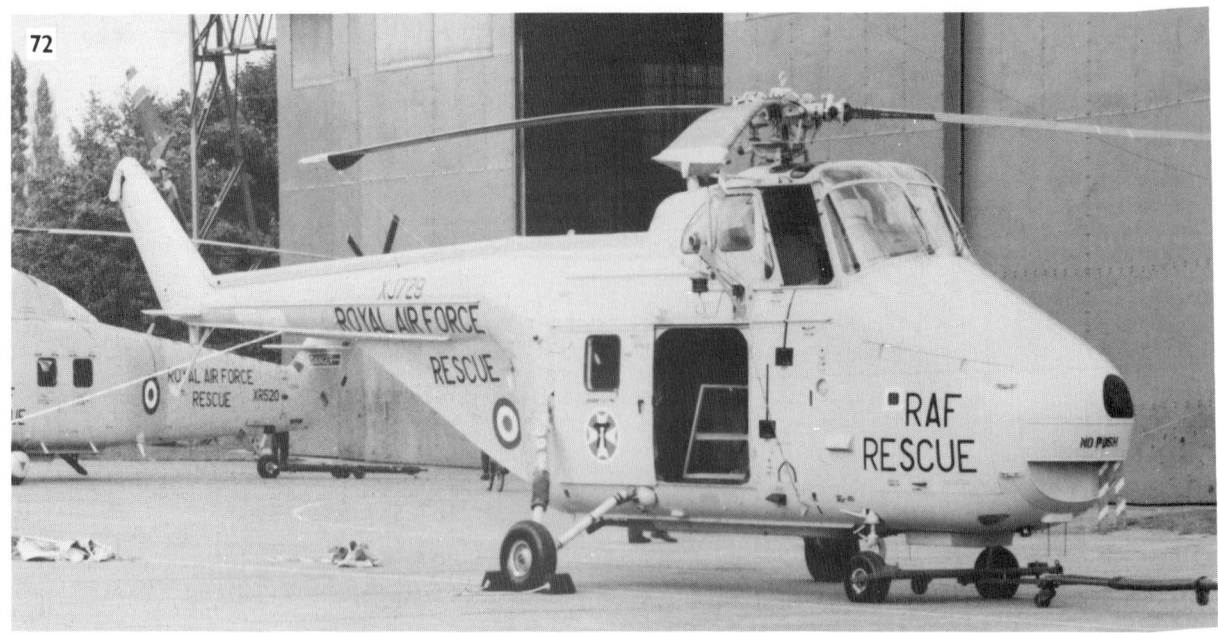

72. Once used in large numbers by the Royal Navy and RAF, the Whirlwind was yet another Westland derivative of an early Sikorsky design. A few examples are still operated by civil concerns, including one former RAF example which retains its service markings. The only remaining military operator is thought to be Brazil, which may still fly a handful of Whirlwinds. XJ729 is a beautifully restored example of an RAF HAR.10 once used for search and rescue but now preserved at RAF Finningley in the SAR Engineering headquarters. [*Tim Laming*]

73. Not quite an aeroplane, not quite a helicopter. The Bell/Boeing V-22 Osprey is expected to be the first serious military application of the tilt-rotor concept, whereby the machine can take off vertically like a helicopter, and then tilt the rotors forward (effectively changing the rotors into propellers) to begin conventional flight. Shown here is 3914 during its first flight at Bell's Arlington factory in Texas. Budget constraints have forced the US Government to vary its support for the programme, and whether the V-22 eventually enters service with the USMC remains to be seen. However, it does look as if the V-22 will have a future – somewhere. [*Bell*]